HOW TO BALANCE LIVE SOUND MIXING FOR ENGINEERS

A COMPREHENSIVE GUIDE ON HOW TO BALANCE LIVE SOUND MIXING FOR ENGINEERS: TECHNIQUES, STRATEGIES, AND BEST PRACTICES FOR ACHIEVING OPTIMAL CLARITY

Timbreer kancher

TABLE OF CONTENT

INTRODUCTION

PURPOSE OF THE BOOK

Live sound blending is an art and a technology, requiring a deep information of each technical standards and innovative instinct. The intention of this e book is to provide a complete guide for sound engineers, from the ones just beginning in the field to seasoned experts in search of to refine their skills. Whether you are running in small golf equipment, huge concert halls, or outside festivals, this e book will equip you with the know how and techniques necessary to supply awesome sound constantly.

In modern day speedy paced world, stay occasions have become a cornerstone of the leisure industry. Audiences assume impeccable sound best, and the function of the live sound engineer is extra critical than ever. This book will help you meet and exceed those expectations by using masking a huge range of topics, from the basics of acoustics to superior blending strategies.

TARGET AUDIENCE

This book is designed for:

 Aspiring Sound Engineers: Those who are new to the sphere and want to construct a solid foundation in live sound mixing.

 Experienced Sound Engineers: Professionals looking to enlarge their knowledge and stay modern with the brand new strategies and technology.

 Musicians and Performers: Artists who need to recognize the technical components of stay sound to improve their performances and communique with sound engineers.

 Event Coordinators and Producers: Individuals responsible for organizing stay events who want a higher knowledge of sound necessities and the paintings of sound engineers.

IMPORTANCE OF LIVE SOUND MIXING

Live sound blending is a vital issue of any live overall performance, influencing the general enjoy of the audience and the fulfillment of the event.

1. Audience Experience

The great of live sound mixing at once impacts the target audience's leisure and engagement. Clear, well balanced sound guarantees that every detail of the overall performance is heard, from the diffused nuances of a vocalist to the powerful beats of a drummer. Good sound blending can rework a median overall performance into an unforgettable experience, even as bad blending can detract from even the maximum proficient performers.

2. Performer Satisfaction

Performers rely heavily on sound engineers to offer them with the pleasant viable sound on stage. Properly blended monitors and inear structures allow musicians to pay attention themselves and each different in reality, leading to better performances. When performers are assured in their sound, they could awareness greater on their artistry and less on technical problems.

3. Technical Precision

Live sound mixing calls for a excessive level of technical expertise. Engineers have to understand the intricacies of sound equipment, signal flow, and acoustics to create a balanced blend. This technical precision is crucial for preventing troubles like remarks, distortion, and segment problems, that may disrupt the overall performance and harm gadget.

4. Adaptability and Problem Solving

Live activities are dynamic, with many variables which can change at a moment's note. A skilled sound engineer must be capable of adapt quickly to sudden conditions, consisting of system screw ups, changes within the overall performance lineup, or various acoustics in exclusive venues. Effective trouble solving guarantees that the display is going on easily, irrespective of any demanding situations that get up.

5. Professional Reputation

For sound engineers, the fine in their work can considerably effect their expert reputation. Consistently delivering wonderful sound mixes builds believe with performers, occasion organizers, and audiences. This recognition can lead to greater job possibilities and career development in the aggressive discipline of stay sound.

DIFFERENCES BETWEEN LIVE AND STUDIO SOUND MIXING

While both stay and studio sound blending aim to gain top notch audio, they fluctuate notably of their environments, techniques, and desires. Understanding those variations is critical for sound engineers running in either or each settings.

1. Environment

Live Sound Mixing:

Takes area in diverse dynamic and frequently unpredictable environments, inclusive of concert halls, golf equipment, outside venues, and theaters.

Sound engineers have to account for the acoustics of the venue, that can exchange with the presence of an audience.

External elements which includes weather situations in outdoor settings can effect sound quality and system performance.

Studio Sound Mixing:

- Occurs in a managed, acoustically handled surroundings designed to reduce outside noise and reflections.
- The constant and predictable acoustic properties of the studio permit for precise and unique blending.

2. Time Constraints

Live Sound Mixing:

- Mixing is finished in real time for the duration of the overall performance, requiring short decision making and instant adjustments.
- There is little to no opportunity for publish manufacturing modifying, making it critical to get matters proper immediate.

Studio Sound Mixing:

- Allows for considerable time to test, refine, and best the mix.

- Engineers can use a couple of takes, layers, and publish production strategies to achieve the preferred sound.

3. Equipment and Technology

Live Sound Mixing:

- Utilizes transportable and strong device designed to withstand the rigors of transportation and setup in exceptional venues.

- Mixing consoles are regularly virtual with capabilities tailor made for stay sound, consisting of scene don't forget and real time consequences processing.

- Sound reinforcement structures, which includes PA systems and stage monitors, are essential for amplifying sound to the target market and performers.

Studio Sound Mixing:

- Employs high fidelity, often extra sensitive gadget optimized for specified sound replica.

- Mixing consoles in studios can be analog or virtual, with a focal point on enormous processing skills and automation.
- Studio video display units are designed for correct, flat frequency reaction to make sure particular mixing.

4. Feedback and Monitoring

Live Sound Mixing:

- Engineers have to be vigilant about preventing and coping with feedback, that could arise without difficulty in stay settings due to the interaction between microphones and speakers.
- Monitor mixes are critical for performers to pay attention themselves and each different on level, often requiring separate mixes for one of a kind performers.

Studio Sound Mixing:

- Feedback is typically no longer an difficulty in view that monitoring is performed via headphones or studio monitors in a controlled environment.

- Engineers can attention on developing a balanced and polished mix with out the instant worries of remarks.

5. Collaboration and Communication

Live Sound Mixing:

- Involves real time communication with performers, stage managers, and other crew participants to address any troubles and ensure easy transitions throughout the overall performance.

- Sound tests and rehearsals are crucial for putting in and quality tuning the combination before the real overall performance.

Studio Sound Mixing:

- Allows for extra deliberate and exact communication with artists and manufacturers to reap the preferred sound.
- Collaboration may be greater iterative, with the possibility for more than one revisions and remarks sessions.

6. Goals and Outcomes

Live Sound Mixing:

- The primary purpose is to deliver an immersive and enjoyable revel in for the audience, ensuring readability, stability, and effect.
- The blend need to adapt to the stay performance's strength and spontaneity, improving the artist's reference to the audience.

Studio Sound Mixing:

- Aims to create a cultured, expert recording that sounds awesome on diverse playback structures, from high give up audio system to consumer headphones.

- The recognition is on achieving an in depth, nuanced blend that can stand the check of time and meet enterprise standards for best.

CHAPTER ONE:

FUNDAMENTALS OF SOUND

BASIC SOUND PRINCIPLES

SOUND WAVES AND FREQUENCIES

Understanding sound waves and frequencies is essential to mastering live sound blending. These principles form the premise of ways we perceive sound and the way sound behaves in one of a kind environments.

Sound Waves

1. Nature of Sound Waves

Definition: Sound waves are longitudinal waves that journey via a medium (which include air, water, or solid materials) due to the vibration of debris. These waves are created by vibrating objects, which motive the surrounding medium to vibrate as well.

Propagation: Sound waves propagate by means of compressing and rarefying the particles in the medium. This compression (high strain area) and rarefaction (low stress location) procedure moves the wave ahead.

2. Characteristics of Sound Waves

Amplitude: This is the height of the sound wave and determines the quantity or loudness of the sound. Higher amplitude manner a louder sound, even as lower amplitude approach a softer sound.

Wavelength: The distance among consecutive points in segment on the wave (e.G., compression or two rarefaction's). Wavelength is inversely proportional to frequency.

Frequency: The number of complete wave cycles that bypass a point in one 2nd, measured in Hertz (Hz). Higher frequency method greater cycles per second and corresponds to a better pitch, while lower frequency approach fewer cycles consistent with 2d and corresponds to a lower pitch.

Velocity: The pace at which the sound wave travels via a medium. In air, sound travels at approximately 343 meters in step with second (1,a hundred twenty five ft in step with 2nd) at room temperature.

FREQUENCIES

1. Frequency Range and Perception

Human Hearing Range: The standard range of human hearing extends from about 20 Hz to twenty,000 Hz (20 kHz). Sounds below 20 Hz are called infra sound, and sounds above 20 kHz are known as ultrasound.

Pitch: The perception of frequency through the human ear. Higher frequencies are perceived as better pitches, and lower frequencies are perceived as decrease pitches.

2. Frequency Bands

Sub Bass (2060 Hz): The lowest frequencies, regularly felt extra than heard. These upload power and weight to music.

Bass (60250 Hz): Fundamental frequencies for many musical units and the human voice. This range gives a experience of fullness and heat.

Low Mid range (250500 Hz): Contains the decrease harmonics of maximum contraptions and vocals. Too tons strength on this range could make the sound muddy.

Mid range (500 Hz2 kHz): The center frequencies for the human ear's sensitivity. Crucial for the intelligibility of speech and the presence of devices.

DECIBELS AND SOUND PRESSURE LEVELS

Decibels (dB) and sound pressure stages (SPL) are important standards in know how and managing sound. They help quantify and describe the depth of sound, permitting sound engineers to make informed choices about extent, stability, and universal sound great.

DECIBELS (DB)

1. Definition and Measurement

Decibel (dB): A logarithmic unit used to measure the ratio of a selected quantity to a reference level. In audio, it frequently measures sound depth, power, or strain.

Logarithmic Scale: Because the human ear perceives sound depth logarithmic ally, the decibel scale is likewise logarithmic. This way each 10 dB boom represents a tenfold boom in depth.

For example, a valid at 60 dB is ten instances greater severe than a legitimate at 50 dB.

2. Types of dB Measurements

dB SPL (Sound Pressure Level): Measures the stress of a sound relative to a reference stage of 20 micropascals (μPa), the everyday threshold of human listening to. It quantifies how loud a valid is inside the environment.

Reference Level: zero dB SPL = 20 μPa

dBu and dBV: Measurements utilized in audio device to indicate voltage stages.

dBu is referenced to 0.775 volts.

dBV is referenced to 1 volt.

3. Applications in Live Sound

Volume Control: Adjusting the extent of sound resources (e.G., microphones, instruments) to obtain a balanced blend with out distortion.

Dynamic Range: The distinction between the quietest and loudest components of a performance, measured in dB. Managing dynamic variety guarantees readability and stops signal clipping.

Gain Staging: Setting appropriate input and output degrees in the course of the sign chain to preserve foremost signalization ratio and keep away from distortion.

HUMAN HEARING RANGE

The human hearing range is a crucial factor of sound engineering, because it defines the frequencies that human beings can understand and respond to. Understanding this variety facilitates sound engineers tailor audio reports which can be clean, balanced, and enjoyable for listeners.

HUMAN HEARING FREQUENCY RANGE

1. Frequency Spectrum

Range: The regular variety of human hearing extends from about 20 Hz to 20,000 Hz (20 kHz).

Low Frequencies (20 Hz 250 Hz): Includes sub bass and bass frequencies. These are felt extra than heard and offer the foundation and fullness to track and sound consequences.

Mid range Frequencies (250 Hz four kHz): This variety is vital for the readability of speech and the presence of maximum musical gadgets. The human ear is maximum sensitive to frequencies in this range, particularly around 24 kHz, that is important for speech intelligibility.

High Frequencies (4 kHz 20 kHz): Include top harmonics and overtones that upload brightness and element to sounds. Frequencies above 10 kHz make contributions to the experience of air and area in a mix.

2. Sensitivity and Perception

Equal Loudness Contours: The human ear does now not understand all frequencies similarly at the identical volume degree. These contours, called Fletcher Munson curves, show that the ear is most sensitive to frequencies among 2 kHz and 5 kHz. Lower and higher frequencies need to be louder to be perceived as similarly loud.

Age and Hearing Loss: The capability to listen high frequencies diminishes with age and publicity to loud sounds. This circumstance, called presbycusis, commonly impacts frequencies above 1215 kHz first.

PRACTICAL IMPLICATIONS FOR SOUND ENGINEERS

1. Equalization (EQ)

Balancing the Mix: Understanding the human listening to range allows engineers use EQ to stability different factors of a mix, making sure that all frequencies are accurately represented with out overpowering each different.

Compensating for Sensitivity: Engineers could make modifications to atone for the ear's sensitivity to mid range frequencies and the relative insensitivity to excessive low and excessive frequencies.

2. Monitoring and Playback

Choosing Monitors: Selecting correct studio video display units that provide a flat frequency response across the whole human listening to range is vital for specific mixing.

Hearing Protection: Using ear protection in loud environments to prevent hearing loss, ensuring long term capability to mix successfully.

3. Dynamic Range

Maintaining Clarity: Ensuring the dynamic variety of a mixture is suitable for the intended playback environment, thinking about the ear's varying sensitivity to extraordinary frequencies.

Strategies for special venues (indoor vs. Out of doors)

Strategies for Different Venues (Indoor vs. Outdoor)

Sound engineers face unique challenges and opportunities whilst mixing for special forms of venues. Strategies vary drastically among indoor and out of doors settings due to variations in acoustics, ambient noise, and environmental factors.

INDOOR VENUES

1. Understanding Room Acoustics

Reverberation: Indoor spaces often have reflective surfaces that purpose reverberation. Engineers should consider the reverberation time and the way it impacts readability.

Strategy: Use acoustic remedy (e.G., panels, bass traps) to lessen unwanted reflections. Adjust EQ to control frequencies that make contributions to muddiness or excessive reverberation.

Room Modes: Standing waves can cause sure frequencies to be amplified or canceled out in particular areas of the room.

Strategy: Identify problematic frequencies the usage of a spectrum analyzer and practice precise EQ cuts. Position speakers and subwoofer to reduce the effect of standing waves.

2. Speaker Placement and Coverage

Speaker Positioning: Ensure even sound insurance at some point of the venue to avoid lifeless spots and overly loud regions.

Strategy: Use a couple of speaker arrays or delay speakers to provide steady insurance. Aim audio system to direct sound where the target audience is positioned and faraway from reflective surfaces.

Monitor Placement: Proper placement of stage video display units to make sure performers can listen themselves without inflicting comments.

Strategy: Use inear video display units or function wedge monitors cautiously, angling them to reduce feedback.

3. Managing Feedback

Feedback Prevention: Indoor environments can be susceptible to remarks due to near proximity of microphones and audio system.

Strategy: Use directional microphones to reduce pickup of undesirable sounds. Apply notch filters to problematic frequencies and use remarks suppression gear if essential.

4. Ambient Noise Control

Minimizing External Noise: Indoor venues can once in a while have external noise assets (e.G., HVAC systems, audience chatter).

Strategy: Use noise gates to lessen the effect of ambient noise on open microphones. Position microphones to minimize pickup of unwanted sounds.

OUTDOOR VENUES

1. Adapting to Open Air Acoustics

Lack of Reflections: Outdoor venues generally have minimum reflective surfaces, resulting in less herbal reverb.

Strategy: Use synthetic reverb and postpone consequences to create a experience of space and depth. Adjust the combination to make certain readability and definition inside the absence of natural reflections.

Sound Dispersion: Sound disperses greater freely in outdoor settings, doubtlessly leading to choppy coverage.

Strategy: Employ line array structures to acquire controlled dispersion and regular insurance. Use delay towers for large venues to keep synchronization and sound pleasant over distance.

2. Weather Considerations

Weather Impact: Outdoor occasions are issue to weather conditions inclusive of wind, rain, and temperature changes, that may affect sound pleasant and gadget performance.

Strategy: Use climate resistant gadget and covers to protect equipment from the elements. Monitor climate forecasts and feature contingency plans for unfavorable situations.

3. Managing Ambient Noise

Environmental Noise: Outdoor venues often have higher ambient noise tiers from site visitors, wind, and different assets.

Strategy: Use directional microphones to cognizance on the meant sound sources. Apply excessive bypass filters to lessen low frequency rumble from wind and other environmental noises.

4. Power and Signal Distribution

Power Supply: Ensure solid and enough power supply for all device, as outside venues can also have more hard energy distribution.

Strategy: Use energy conditioners and uninterruptible power components (UPS) to guard device and make certain uninterrupted performance.

Signal Integrity: Long cable runs in out of doors settings can result in signal loss and interference.

Strategy: Use brilliant cables and, if necessary, energetic DI boxes to keep sign integrity over lengthy distances.

5. Audience and Stage Layout

Audience Placement: Consider the layout of the audience vicinity and the way it affects sound coverage and nice.

Strategy: Use speaker zoning to cater to different audience sections. Ensure the front of house (FOH) and screen mixes are optimized for each target market and performers.

Stage Setup: Plan the stage layout to facilitate smooth sound distribution and effective monitoring.

Strategy: Position instruments and monitors to minimize bleed and make sure clear tracking for performers.

CHAPTER TWO: ESSENTIAL EQUIPMENT

MICROPHONES

TYPES (DYNAMIC, CONDENSER, RIBBON)

Understanding the varieties of microphones is vital for sound engineers, as every kind has its unique traits and programs.

Dynamic Microphones

1. Overview

Construction: Dynamic microphones use a diaphragm connected to a coil of wire, placed in the magnetic subject of a magnet. When sound waves hit the diaphragm, it movements the coil, producing an electrical modern day.

Durability: They are sturdy and may face up to high sound strain levels (SPL), making them appropriate for stay sound environments.

2. Characteristics

Frequency Response: Generally less sensitive to excessive frequencies, with a frequency reaction tailored to deal with loud assets.

Sensitivity: Lower sensitivity in comparison to condenser and ribbon microphones, which makes them much less prone to choosing up ambient noise and comments.

Applications: Ideal for miking loud sound resources inclusive of guitar amplifiers, drums (in particular snare and kick drums), and stay vocals. Commonly used dynamic microphones encompass the Shure SM58 and SM57.

3. Advantages and Disadvantages

Advantages: Durable, less costly, proof against moisture, and capable of managing excessive SPL.

Disadvantages: Limited frequency reaction and sensitivity in comparison to different types, resulting in less detail and clarity.

CONDENSER MICROPHONES

1. Overview

Construction: Condenser microphones use a diaphragm placed near a backplane, forming a capacitor. When sound waves hit the diaphragm, the distance among the diaphragm and the back plate modifications, changing the capacitance and generating an electrical sign.

Power Requirement: They require an external strength supply, generally phantom strength (48V), to operate.

2. Characteristics

Frequency Response: Wide and flat frequency reaction, able to taking pictures excessive frequency detail and transients correctly.

Sensitivity: High sensitivity, which permits them to capture subtle nuances and quieter sounds.

Applications: Suitable for studio recording of vocals, acoustic instruments, and overhead miking of drums. Popular condenser microphones include the Neumann U87 and the Audio Technica AT2020.

3. Advantages and Disadvantages

Advantages: High sensitivity, wide frequency response, specified sound seize, and flexibility.

Disadvantages: More fragile and touchy to moisture, managing noise, and comments. Generally greater costly and calls for phantom energy.

RIBBON MICROPHONES

1. Overview

Construction: Ribbon microphones use a skinny metallic ribbon suspended between the poles of a magnet. Sound waves motive the ribbon to transport, generating an electrical sign.

Vintage Appeal: Known for their warm, herbal sound and vintage character, they have been usually used in the early days of broadcasting and recording.

2. Characteristics

Frequency Response: Smooth and herbal frequency reaction, with a selected emphasis on the mid range. Often less sensitive to very high frequencies, which contributes to their heat sound.

Sensitivity: High sensitivity however can be sensitive and susceptible to harm from high SPL and plosive sounds.

Applications: Excellent for recording vocals, strings, brass units, and as room mics. They are prized for his or her herbal sound reproduction and smoothness. Classic examples consist of the Royer R121 and the AEA R84.

3. Advantages and Disadvantages

Advantages: Warm, natural sound with a smooth frequency reaction. Ideal for shooting the man or woman and nuance of the supply.

Disadvantages: Fragile, vulnerable to harm from excessive SPL and wind blasts, sensitive to handling noise, and typically greater costly.

CHOOSING THE RIGHT MICROPHONE

When deciding on a microphone, remember the following factors:

Sound Source: Match the microphone type to the sound supply and its characteristics (e.G., dynamic for loud sources, condenser for special recording, ribbon for natural sound).

Environment: Consider the environment (stay vs. Studio) and capability troubles which includes ambient noise, remarks, and managing noise.

 Budget: Balance the need for exceptional with price range constraints, as microphones range broadly in price.

MIXING CONSOLES

Mixing consoles, additionally referred to as audio mixers or soundboards, are critical to the manner of live sound mixing and studio recording. They allow sound engineers to manipulate and manipulate audio alerts from numerous sources to create a balanced and cohesive mix.

 COMPONENTS OF A MIXING CONSOLE

1. Input Channels

 Microphone Preamps: Amplify weak microphone indicators to line stage for processing.

Line Inputs: Receive signals from instruments, playback devices, or different sources.

Phantom Power: Provides 48V power to condenser microphones and lively DI containers.

2. Channel Strip

EQ (Equalization): Adjusts the frequency response of the signal to beautify or attenuate particular frequencies.

Auxiliary Sends (Aux Sends): Routes signals to outside effects processors or reveal mixes.

Pan/Balance Control: Adjusts the location of the signal inside the stereo discipline (left proper).

three. Faders

Main Faders: Control the overall volume (degree) of each channel inside the main blend.

Subgroup Faders: Combine multiple channels into subgroups for less complicated manipulate and processing.

Monitor Sends: Send alerts to degree monitors for performers to pay attention themselves and every different.

4. Master Section

Master Fader: Controls the overall output degree of the principle blend dispatched to the audience.

Master EQ: Global EQ for the entire mix, used to alter overall tonal stability.

Subgroup Outputs: Provides outputs for subgroups that may be used for added processing or routing.

5. Monitoring and Cueing

Headphone Outputs: Allows sound engineers and performers to display specific channels or mixes.

Solo (PFL/AFL): Prefader and after fader solo alternatives for separating and monitoring person indicators.

Metering: Visual signs (LED meters) display signal ranges to save you clipping and make sure choicest benefit staging.

ANALOG VS. DIGITAL MIXERS

Choosing among analog and virtual mixers is a large selection for sound engineers, impacting workflow, sound satisfactory, flexibility, and usual overall performance.

ANALOG MIXERS

1. Operation

Signal Processing: Analog mixers use bodily additives (resistors, capacitors, operational amplifiers) to manner audio signals.

Control Interface: Physical knobs, sliders, and switches offer tactile manipulate over every channel and parameter.

Sound Character: Analog circuitry can impart a heat, natural sound first rate, frequently favored in certain musical genres and for stay performances.

2. Advantages

Ease of Use: Intuitive interface with on the spot palmson manage, making it easier for engineers to analyze and operate.

Signal Integrity: Direct, nonstop sign route without the ability latency brought through digital processing.

Durability: Generally robust and less susceptible to virtual system defects or software disasters.

3. Limitations

Flexibility: Limited in terms of re callable presets and automated blending capabilities found in digital consoles.

Size and Weight: Larger and heavier due to the bodily components required for signal processing.

Integration: May require additional outboard gear for results processing and recording talents.

DIGITAL MIXERS

1. Operation

Signal Processing: Digital mixers convert analog signals into digital information for processing the usage of DSP (Digital Signal Processing) chips.

Control Interface: Touchscreens, encoders, and virtual shows provide complete manage over all parameters, frequently with customization layouts.

Sound Quality: High decision audio processing with the potential to apply unique EQ, dynamics, and effects.

2. Advantages

Flexibility and Versatility: Extensive routing options, re callable presets, and digital outcomes processing decorate workflow efficiency.

Integration: Seamless integration with DAWs (Digital Audio Workstations), recording systems, and networked audio setups.

Compactness: Smaller footprint and lighter weight in comparison to analog opposite numbers, suitable for transportable setups.

3. Features

Automation: Automated mixing functions, scene don't forget, and motorized faders simplify complex blending obligations and allow particular changes.

Effects and Processing: Builtin effects processors, multi band EQs, dynamics processing, and extensive routing talents.

Networking: Support for virtual audio networking protocols (e.G., Dante, AVB) for scalable, networked audio structures.

CONSIDERATIONS FOR CHOOSING BETWEEN ANALOG AND DIGITAL MIXERS

1. Workflow Preferences

Analog: Ideal for engineers who choose tactile control and a honest interface.

Digital: Suited for folks who price flexibility, automation, and integration with contemporary digital audio workflows.

2. Sound Preference

Analog: Desired for its feature warmth and harmonic distortion that may decorate sure genres of music.

Digital: Offers pristine sound nice, particular processing, and the ability to replicate clean, transparent audio.

3. Application and Environment

Live Sound: Digital mixers offer blessings in complex routing, scene don't forget, and far flung control abilities.

Studio Recording: Both analog and virtual mixers have their vicinity, relying at the desired sound and workflow possibilities.

4. Budget and Future Proofing

Cost: Analog mixers can be greater cost effective in advance, even as virtual mixers can also require a better initial funding.

Longevity: Digital mixers often offer destiny proofing thru software program updates and expand ability, while analog consoles might also have restrained upgrade paths.

EQUALIZERS, COMPRESSORS, AND LIMITERS

Equalizers (EQ), compressors, and limiters are important tools in audio processing, used by sound engineers to shape and control the dynamics, tonal stability, and universal sound fine of audio alerts.

Equalizers (EQ)

1. Purpose

Frequency Adjustment: Equalizers modify the frequency reaction of audio alerts with the aid of boosting or slicing unique frequency stages.

Tonal Balance: Used to balance the mix by using emphasizing or attenuating certain frequencies to obtain clarity and concord.

Corrective and Creative: Can be used to correct intricate frequencies (e.G., room resonance) or creatively form the sound (e.G., including presence or warmth).

2. Types

Graphic EQ: Consists of multiple constant frequency bands with character sliders for every band (e.G., 31band or 15band EQ).

Parametric EQ: Offers more manipulate with adjustable frequency bands (center frequency, bandwidth/Q, advantage).

Shelving EQ: Boosts or cuts all frequencies above or below a sure point (shelf).

3. Applications

Mixing: Used to carve out area for unique units and vocals within the mix, prevent masking, and enhance readability.

Mastering: Fine tunes the overall tonal stability of the mix for consistency across extraordinary playback structures.

Live Sound: Manages comments, adjusts for room acoustics, and tailors the sound to the venue.

COMPRESSORS

1. Purpose

Dynamic Range Control: Compressors lessen the dynamic range of audio indicators with the aid of attenuating louder components, thereby growing perceived loudness and controlling peaks.

Envelope Shaping: Adjusts the assault, launch, ratio, and threshold to shape the envelope of the sound.

Glue and Cohesion: Used to attach together tracks in a combination, smooth out inconsistencies, and add preserve to devices.

2. Parameters

Threshold: Determines the extent at which compression starts to apply.

Ratio: Sets the quantity of benefit reduction applied as soon as the signal exceeds the threshold (e.G., 4:1 ratio method for each 4 dB above the brink, only 1 dB passes thru).

Attack: Controls how quickly the compressor responds to indicators exceeding the brink.

Release: Determines how quickly the compressor stops attenuating the signal after it falls below the threshold.

3. Applications

Vocal Compression: Smooths out vocal dynamics, ensuring constant levels with out dropping natural expression.

Drum Compression: Controls drum transients, enhances punch, and adds sustain to drum sounds.

Mix Bus Compression: Glues collectively the general blend, enhances perceived loudness, and ensures a cohesive sound.

LIMITERS

1. Purpose

Peak Control: Limiters are intense compressors with a excessive ratio (frequently 10:1 or more) used to save you signal peaks from exceeding a fixed degree (threshold).

Protection: Prevents clipping and distortion by using forcefully limiting the most output level of audio alerts.

Transient Management: Controls sudden, brief peaks without affecting the overall dynamics excessively.

2. Parameters

Threshold: Sets the most output degree at which limiting begins to prevent peaks.

Ratio: Typically very high, making sure any sign exceeding the threshold is significantly attenuated.

Release: Determines how quickly the limiter returns to everyday operation after attenuating peaks.

3. Applications

Mastering: Ensures the final mix stays within desired loudness limits with out clipping.

Live Sound: Protects speakers from harm by way of preventing excessively loud alerts.

Broadcast and Streaming: Maintains steady audio levels to conform with broadcast standards and prevent unexpected quantity adjustments.

CHOOSING AND USING THESE TOOLS

- Understanding Signal Flow: Place EQ earlier than compression inside the signal chain to form frequencies earlier than dynamic processing.

- Experimentation: Adjust parameters (threshold, ratio, attack, release) primarily based at the audio cloth and favored effect.

EFFECTS UNITS: REVERB AND DELAY

Effects units such as reverb and put off play crucial roles in audio manufacturing and stay sound engineering, including spatial depth, texture, and ambiance to audio alerts.

REVERB

1. Purpose

Ambiance and Specialization: Reverb simulates the acoustic environment of a space via reflecting sound waves off surfaces, growing a experience of intensity and natural resonance.

Enhancement: Adds richness, warmth, and realism to dry audio indicators, making them sound greater natural and immersive.

Creative Tool: Used creatively to rouse unique moods (e.G., spacious cathedral reverb for epic soundscapes).

2. Types

Room Reverb: Simulates the acoustic residences of numerous room sizes, from small studios to big live performance halls.

Hall Reverb: Emulates the reverberant sound of live performance halls, presenting longer decay instances and a spacious sense.

Plate Reverb: Mimics the sound of early mechanical reverberation gadgets, acknowledged for their easy and dense reverb tails.

Spring Reverb: Uses a coiled metallic spring to create a specific, somewhat "boingy" reverb impact popular in antique guitar amps.

3. Applications

Vocals: Adds intensity and presence to vocal performances, making them sound greater polished and expert.

Instruments: Enhances the spatial excellent of devices, making them sound greater herbal and blended inside a mix.

Mixing and Mastering: Used to vicinity factors in a digital acoustic space, developing a cohesive and immersive listening enjoy.

DELAY

1. Purpose

Echo Effect: Delay repeats the audio sign after a fixed quantity of time, creating awesome echoes which could variety from diffused repetitions to rhythmic patterns.

Spatial Enhancement: Similar to reverb, delay adds depth and size through creating a feel of distance among the original sound and its repetitions.

Creative Tool: Used for rhythmic consequences, doubling vocals or gadgets, and developing spatial illusions.

2. Types

Analog Delay: Uses analog circuitry to create heat, organic repeats with feature degradation over time.

Digital Delay: Offers unique manipulate over postpone time, comments, and modulation effects, frequently with cleaner and more correct repeats.

Tape Delay: Emulates the sound of antique tape echo machines, known for their warm, saturated repeats and modulation outcomes.

Multi Tap Delay: Provides a couple of postpone faucets with impartial timing and remarks settings, permitting complex rhythmic styles and spatial results.

3. Applications

Guitar and Keyboards: Used to create rhythmic results, solos with depth, and ambient textures.

Vocals: Adds spatial enhancement, doubling consequences, and rhythmic vocal patterns.

Production and Mixing: Enhances stereo width, creates motion within a mixture, and adds depth to recorded tracks.

USING EFFECTS UNITS EFFECTIVELY

Parameters: Adjust parameters inclusive of decay time (reverb), put off time, remarks, and wet/dry mix to obtain the preferred impact.

Blend and Balance: Blend the affected sign with the dry signal to preserve clarity and keep away from overwhelming the combination.

Automation: Automate impact parameters through the years to create evolving textures and dynamic modifications.

Monitoring: Listen seriously to how consequences impact the overall blend, making sure they beautify in place of distract from the musical content material.

TYPES OF SPEAKERS: MAINS, SUBS, AND MONITORS

Understanding the exceptional kinds of audio system utilized in audio systems is important for achieving balanced sound reinforcement and tracking in various environments.

MAINS (FULL RANGE SPEAKERS)

1. Purpose

Primary Sound Reinforcement: Mains are designed to reproduce a extensive frequency variety, usually from lows to highs, covering the whole audible spectrum.

Audience Coverage: Positioned in front of residence (FOH), mains deliver sound to the audience region in live overall performance venues.

2. Characteristics

Frequency Response: Full variety capability to breed bass, mid range, and treble frequencies.

Power Handling: High energy coping with to supply sufficient quantity for big venues.

Dispersion: Designed for extensive dispersion to cover a large vicinity frivolously.

3. Applications

Live Sound: Used in live shows, fairs, and events to deliver clear and balanced sound to the audience.

Fixed Installations: Installed in venues including theaters, auditoriums, and homes of worship for permanent sound reinforcement.

Subs (Sub woofers)

1. Purpose

Low Frequency Reinforcement: Subwoofer specialist in reproducing low frequency bass sounds that mains won't take care of properly.

Enhanced Impact: Provide deep bass extension and effect for music genres like EDM, hip hop, and cinematic soundtracks.

2. Characteristics

Frequency Range: Typically reproduce frequencies underneath one hundred Hz, specializing in sub bass and bass frequencies.

Enclosure Types: Available in various configurations such as bass reflex (ported), sealed, and band pass designs.

Power and Efficiency: High energy handling to breed low frequencies at excessive SPLs correctly.

3. Applications

Live Sound: Augment mains in massive venues to offer deep, effective bass reinforcement.

Studio Mixing: Used in recording studios for accurate low frequency monitoring for the duration of tune production and mixing.

MONITORS (STUDIO MONITORS AND STAGE MONITORS)

1. Studio Monitors

Purpose: Designed for important listening and correct sound reproduction in recording studios.

Near field vs. Midfield: Near field video display units are placed close to the listener for exact monitoring, whilst midfield video display units are used for larger manage rooms.

Frequency Response: Balanced frequency response for accurate representation of audio content.

Enclosure Types: Typically active (powered) with builtin amplification and designed to decrease coloration.

2. Stage Monitors (Wedge Monitors)

Purpose: Provide onlevel audio tracking for performers to pay attention themselves and different instruments all through live performances.

Designs: Available as floor wedges (angled in the direction of the performer) or inear monitors for individualized tracking.

Feedback Rejection: Designed to minimize feedback by way of focusing sound in the direction of the performer and far from microphones.

3. Applications

Studio Monitors: Used in recording, mixing, and learning to ensure accurate sound illustration and selection making.

Stage Monitors: Essential for stay performances, permitting performers to listen themselves and hold pitch and timing.

CHOOSING AND USING SPEAKERS EFFECTIVELY

System Matching: Select speakers that supplement every different in terms of frequency reaction and dispersion traits.

Placement: Position audio system efficaciously for superior coverage and sound exceptional, thinking about venue acoustics and audience placement.

Integration: Ensure proper integration with amplifiers, processors, and combining consoles to maximize performance and reliability.

Monitoring: Regularly screen and calibrate speakers to keep consistent sound quality and prevent device fatigue.

AMPLIFIERS AND CROSSOVERS IN AUDIO SYSTEMS

Amplifiers and crossovers are crucial components in audio structures, liable for powering audio system and managing the distribution of audio frequencies to make certain most excellent sound first class and performance.

Amplifiers

1. Purpose

Signal Amplification: Amplifiers growth the amplitude (volume) of audio alerts to force speakers and deliver enough electricity for sound reinforcement.

Signal Integrity: Ensure constancy and readability by means of retaining sign integrity at some stage in the amplification process.

2. Types

Analog Amplifiers: Traditional amplifiers that use analog circuitry (transistors, tubes) to extend signals. They are known for their warm, natural sound traits and are frequently favored in sure audio applications for his or her sonic characteristics.

Digital Amplifiers (Class D): Efficient amplifiers that convert analog audio signals into digital pulses, that are amplified and then transformed again to analog at the output degree. They are lightweight, strength efficient, and appropriate for high strength programs.

3. Characteristics

Power Rating: Rated in watts consistent with channel (WPC), indicating the most electricity output in line with speaker channel at a given impedance (e.G., 100W in step with channel at 8 ohms).

Impedance Matching: Match amplifier output impedance with speaker impedance to make sure efficient energy switch and prevent harm to equipment.

4. Applications

Live Sound Reinforcement: Powering mains, subs, and video display units in stay concert venues, theaters, and out of doors activities.

Studio Monitoring: Driving studio video display units for accurate audio monitoring and mixing in recording and publish production studios.

Home Audio Systems: Powering home theater structures, stereo setups, and multi room audio distribution structures.

CROSSOVERS

1. Purpose

Frequency Management: Crossovers divide the audio sign into separate frequency bands (low, mid, excessive) and direct each band to the ideal speaker thing (woofer, mid range, tweeter).

Speaker Protection: Prevents distortion and speaker damage by way of ensuring that each speaker receives handiest the frequencies it's miles designed to handle. Enhanced Sound Quality: Optimizes sound exceptional by permitting each speaker driver to operate inside its most beneficial frequency range.

2. Types

- Active Crossovers: Electronic gadgets that break up the audio sign into separate frequency bands earlier than amplification. They are positioned among the audio supply (mixer, preamp) and multiple electricity amplifiers driving character speaker components (e.G., woofers, tweeters).

CHAPTER THREE: SIGNAL FLOW AND ROUTING

UNDERSTANDING SIGNAL FLOW

INPUT TO OUTPUT REVIEW

Input to Output Overview of Audio System

1. Source/Input

Microphones, Instruments, Playback Devices: Audio signals originate from microphones taking pictures sound, contraptions generating electric signals, or playback gadgets (CD players, laptops) sending prerecorded audio.

2. Pre Amplification

Microphone Preamps: Boost vulnerable microphone indicators to line stage for processing.

DI Boxes: Convert instrument signals (e.G., guitar, bass) into balanced indicators appropriate for processing and amplification.

3. Processing

Equalization (EQ): Adjusts frequency reaction to beautify or reduce precise frequency stages.

Compression: Controls dynamic range by using attenuating loud indicators and boosting softer ones.

Effects (Reverb, Delay): Adds spatial intensity, environment, and time based effects to decorate the audio signal.

4. Mixing

Mixing Console: Combines and balances a couple of audio indicators (from microphones, contraptions, and so on.) into a cohesive mix.

Input Channels: Receive indicators from sources, where processing (EQ, dynamics) is carried out.

Aux Sends: Route signals to outside effects processors or monitor mixes.

Pan/Balance Controls: Adjust stereo placement of every channel within the blend.

5. Amplification

Power Amplifiers: Amplify line degree indicators to a degree suitable for using speakers.

Mains: Full variety speakers for reproducing mid range and high frequencies.

Subwoofer: Dedicated to low frequency replica (bass).

Monitors: Provide onstage or studio monitoring for performers or engineers.

6. Crossover

Active Crossovers: Divide the audio signal into separate frequency bands (low, mid, high) and direct every band to the suitable speaker (woofer, mid range, tweeter).

Passive Crossovers: Installed within speaker cabinets to break up the amplified sign into distinct frequency levels for each driving force.

7. Output

Speakers: Convert amplified electrical indicators back into audible sound waves.

Main Speakers: Project sound to the target audience in stay settings.

Subwoofer: Enhance bass frequencies for a fuller sound experience.

Monitors: Provide correct audio comments for performers (onstage video display units) or engineers (studio video display units).

8. Monitoring and Control

Mixing and Mastering: Engineers monitor and adjust the final blend for to prated balance and readability.

Live Sound Reinforcement: Ensures steady sound nice and volume during the venue.

Feedback and Adjustment: Continuous tracking and adjustment of ranges, EQ, and outcomes to maintain preferred sound first rate and save you problems like feedback.

DESIGNING EFFECTIVE LEVEL PLOTS

Designing effective stage plots is important for ensuring clean and green stay performances, where every performer and technician knows their placement and device setup.

Steps to Design an Effective Stage Plot

1. Gather Information

Band/Act Details: Obtain data approximately the variety of performers, instruments, and any unique device (e.G., keyboards, drum kits, DJ setups).

Technical Requirements: Know particular technical wishes which include monitor mixes, microphone possibilities, and degree format choices.

2. Create a Layout Sketch

Draw a Stage Map: Sketch a primary layout of the degree, together with dimensions and any fixed features (e.G., drum riser, middle stage region).

Position Markers: Place markers for each performer, indicating their position relative to middle stage or key reference factors.

3. Placement of Instruments and Equipment

- Instruments: Position units including drum kits, amplifiers, keyboards, and every other electronic gadgets in keeping with the degree format.

- Microphones: Mark microphone placements for vocalists, instrumental microphones, and any strong point microphones (e.G., drum mics).

4. Labeling and Marking

- Use Clear Labels: Label each performer's place with their call or device to suggest wherein they should stand or take a seat.

- Color Coding: Use colors or symbols to distinguish between specific styles of system (e.G., monitors, front of house speakers, device lines).

5. Communicate Stage Flow and Movement

- Flow of Movement: Indicate the glide of movement for performers, which include entrances and exits from stage.

- Stage Entry Points: Mark access and exit points on the degree for smooth navigation at some point of performances.

6. Include Technical Details

- Monitor Mixes: Specify display blend requirements for each performer to make sure they are able to pay attention themselves and others virtually.

- Power Requirements: Note strength needs for instruments, amplifiers, and another electronic gadget to ensure sufficient shops are available.

CHAPTER FOUR: MIXING TECHNIQUES

GAIN STAGING

SETTING PROPER ENTER DEGREES

Setting proper input ranges is vital for accomplishing highest quality sound fine and fending off issues like distortion or noise in audio recordings or stay sound reinforcement.

Steps to Set Proper Input Levels

1. Gain Structure Understanding

Gain Staging: Ensure each element in the sign chain (microphones, contraptions, preamps, mixers) is set to the perfect degree to keep away from signal distortion or noise.

2. Signal Flow Check

Start on the Source: Begin with the audio supply (microphone, instrument) and verify its output stage.

Check Cable Connections: Ensure all cables are securely linked and in accurate condition to avoid sign loss or interference

3. Use Peak Meters

Monitor Peak Levels: Use top meters to your mixer or audio interface to reveal sign degrees.

Target Levels: Aim for sign peaks to attain round 12 dB to 6 dB on common, with occasional peaks reaching close to zero dB (however not exceeding zero dB to keep away from clipping).

4. Adjust Input Gain

Preamp or Mixer Gain: Adjust the gain manipulate for your preamp or mixer channel for every enter source.

Optimal Levels: Increase the advantage until the sign peaks are within the favored variety for your height meters.

5. Avoid Clipping

Monitor Peaks: Watch for clipping signs for your mixer or audio interface, which indicate that the sign has handed the most level and is distorting.

Reduce Gain: If clipping takes place, lessen the input advantage barely till the signal peaks now not clip.

6. Check Signal to Noise Ratio

Background Noise: Listen for any unwanted heritage noise that may grow to be greater substantive whilst the advantage is too excessive.

Balance: Find a stability where the sign is robust enough to be clear and gift with out amplifying background noise excessively.

AVOIDING CLIPPING AND DISTORTION

Avoiding clipping and distortion in audio signals is crucial for retaining easy, top notch sound in both recording and stay sound environments. Strategies to Avoid Clipping and Distortion

1. Set Proper Input Levels

- Use Peak Meters: Monitor enter ranges the usage of peak meters for your mixer, audio interface, or recording software program.

- Target Levels: Aim for signal peaks to commonly hit round 12 dB to 6 dB on average, with occasional peaks accomplishing near 0 dB.

- Avoid 0 dB: Ensure indicators do not continuously reach or exceed zero dB to save you clipping.

2. Adjust Gain Carefully

Preamp/Mixer Gain: Set input advantage levels appropriately for every source (microphones, devices).

Start Low: Begin with decrease gain settings and progressively increase till the sign reaches most useful tiers without clipping.

Listen for Distortion: If the sign sounds distorted or harsh, lessen the gain until the distortion disappears.

3. Use Compression Wisely

Dynamic Range Control: Apply compression to govern peaks and prevent sudden spikes in extent that would result in clipping.

Threshold Setting: Set the compressor threshold just above the common stage of the signal to capture peaks with out overly compressing the whole sign.

4. Monitor Signal Flow

Check Cable Quality: Ensure cables are in excellent situation and securely linked to save you signal loss or interference.

Verify Connections: Double test all connections from source to recording device or mixer to make sure right sign flow.

5. Avoid Over driving Processors

Avoid Overloading Effects: Ensure outcomes processors (EQs, reverbs) are not pushed too difficult, that could introduce unwanted distortion.

Monitor Output Levels: Keep an eye fixed on output levels from processors to ensure they're not causing the sign to clip downstream.

EQUALIZATION (EQ)

Equalization (EQ) is a fundamental device in audio processing used to adjust the frequency balance of audio signals.

Equalization (EQ)

1. Purpose Frequency Adjustment: EQ allows you to reinforce or cut specific frequency tiers inside an audio sign.

Tonal Balance: Helps to form the general sound by means of emphasizing or attenuating certain frequencies.

Problem Solving: Corrects frequency imbalances as a result of room acoustics, microphone traits, or instrument timbres.

2. Types of EQ

Graphic EQ: Consists of a couple of fixed frequency bands with person sliders, generally starting from five to 31 bands. Each band covers a particular range of frequencies.

- Parametric EQ: Offers extra flexibility with adjustable parameters:

- Frequency: Sets the middle frequency of the EQ band.

- Gain: Adjusts the amplitude (improve or cut) of the chosen frequency.

- Bandwidth (Q): Controls the width of the frequency band affected by the EQ adjustment.

- Shelving EQ: Adjusts all frequencies above or underneath a positive point (shelf frequency). High shelf (boosts or cuts frequencies above the shelf factor) and coffee shelf (boosts or cuts frequencies beneath the shelf point) are not unusual.

3. Parameters

- Frequency: Determines which frequency band is tormented by the EQ adjustment.

- Gain: Sets the amount of raise (fine gain) or reduce (terrible benefit) implemented to the chosen frequency band.

- Q Factor (Bandwidth): Controls the width of the affected frequency variety. A better Q narrows the bandwidth, affecting fewer frequencies around the center factor.

4. Applications

Mixing and Mastering: EQ is used to stability the tonal characteristics of man or woman tracks and attain clarity and separation inside the blend.

Live Sound Reinforcement: Helps to make amends for room acoustics, comments problems, and beautify the clarity and intelligibility of vocals and instruments.

Instrumental EQ: Shapes the sound of individual instruments to in shape inside the usual blend, emphasizing sure frequencies to decorate presence or warm temperature.

5. Techniques

Cut Before Boost: Generally, it's better to cut elaborate frequencies (e.G., resonances, history noise) earlier than boosting frequencies.

Subtraction EQ: Reduces frequencies that muddy the combination or motive overlaying, enhancing readability with out including pointless boosts.

Surgical EQ: Precisely targets specific hassle frequencies or resonances that detract from the overall sound firstrate.

Frequency tiers and their characteristics

UNDERSTANDING FREQUENCY LEVELS AND THEIR CHARACTERISTICS IS IMPORTANT FOR AUDIO

engineers and producers to correctly control and shape sound.

FREQUENCY RANGES AND CHARACTERISTICS

1. Sub Bass (20 Hz 60 Hz)

Characteristics: Very low frequencies which are felt more than heard, presenting a foundation for bass heavy music genres.

Instruments: Kick drums, bass synthesizers, and sub bass elements in electronic song.

2. Bass (60 Hz 250 Hz)

Characteristics: Fundamental frequencies that add warm temperature, frame, and strength to audio alerts.

Instruments: Bass guitars, floor toms, decrease sign in of vocals, and low pitched contraptions.

three. Low Mid range (250 Hz 500 Hz)

Characteristics: Adds fullness and thickness to sounds, regularly influencing the perceived warm temperature and intensity of audio.

Instruments: Snare drums, acoustic guitars, male vocals, and decrease sign up of pianos.

4. Mid range (500 Hz 2 kHz)

Characteristics: Most sensitive range for human hearing, where maximum harmonic content of musical units and vocals live.

Instruments: Electric guitars, pianos, vocals (both male and girl), and fundamental frequencies of many devices.

5. Upper Mid range (2 kHz 4 kHz)

Characteristics: Adds presence, readability, and articulation to audio indicators, important for intelligibility in speech and vocals.

Instruments: Snare drum attack, electric powered guitars (presence), and consonant sounds in vocals.

6. Presence (4 kHz 6 kHz)

Characteristics: Enhances definition and clarity, making sounds extra "gift" and slicing thru a mix.

Instruments: Cymbals, higher check in of vocals, and a few harmonic content of string instruments.

COMPRESSION AND ITS PROGRAMS

Compression is a vital audio processing device used in both recording and live sound environments to manipulate the dynamic range of audio indicators.

Compression: Overview and Applications

1. Purpose of Compression

Dynamic Range Control: Reduces the distinction among loud and quiet components of audio indicators, making quieter sounds louder and stopping louder sounds from peaking or distorting.

Volume Consistency: Ensures extra steady quantity levels, enhancing typical perceived loudness and clarity.

Enhanced Presence: Brings out details in audio through lowering peaks and permitting lower degree details to grow to be more audible.

2. Components of Compression

Threshold: Sets the level above which compression starts off evolved to affect the signal. Signals above this threshold are reduced in extent.

Ratio: Determines how an awful lot the sign above the threshold is reduced. For instance, a ratio of 4:1 approach that for every 4 dB the signal exceeds the threshold, simplest 1 dB passes through.

Attack: Determines how quickly the compressor reacts once the signal exceeds the edge. A faster attack time reduces the initial temporary, while a slower assault shall we greater of the temporary thru before compression kicks in.

Release: Sets how quick the compressor returns to everyday (no compression) after the signal falls underneath the threshold. Faster release instances can create a more aggressive pumping impact, even as slower releases are smoother.

3. Applications of Compression

Vocal Compression: Smooths out vocal performances with the aid of lowering peaks and enhancing consistency, making vocals extra present in the mix.

Instrumental Balance: Controls the dynamic range of contraptions like drums, bass guitar, and piano to make certain they sit down nicely within the blend with out overpowering other factors.

Live Sound Reinforcement: Manages dynamic peaks in live performances, making sure a greater even sound and preventing remarks from unexpected quantity adjustments.

Broadcasting and Podcasting: Ensures steady audio degrees for listeners, specifically important in radio, podcasts, and streaming where content material varies widely in quantity.

Mastering: Applies subtle compression throughout a whole mix to glue elements together, decorate loudness, and create a cohesive sound.

CHAPTER FIVE: LIVE MIXING STRATEGIES

BALANCING THE MIX

PRIORITIZING UNITS AND VOCALS

Prioritizing gadgets and vocals in a combination entails balancing their levels and frequencies to make certain readability, effect, and cohesion.

Prioritizing Instruments and Vocals in a Mix

1. Establish a Clear Hierarchy

Identify Key Elements: Determine which contraptions or vocals are crucial to the song's association and emotional impact.

Focus on Lead Elements: Highlight lead vocals, melody devices, or key rhythmic elements that pressure the song.

2. Frequency Management

- EQ Techniques: Use EQ to carve out frequency area for every tool and vocal:

- High bypass Filters: Remove pointless low frequencies from non bass contraptions and vocals.

- Notch Filtering: Address frequency clashes via notching out unique frequencies wherein instruments or vocals overlap.

- Frequency Masking: Address conflicts wherein two elements occupy similar frequency degrees, adjusting EQ to decorate readability.

3. Dynamic Range Control

- Compression: Apply compression to control dynamics and make sure vocals and lead gadgets stay distinguished without overwhelming different elements.

- Vocals: Use mild compression to easy out dynamics, making sure consistent presence without sounding overly processed.
- Instruments: Tailor compression settings to devices, keeping their natural dynamics while maintaining clarity and impact.

4. Panning and Stereo Imaging

Spatial Placement: Place vocals and lead units centrally or barely off center to keep consciousness and readability.

Stereo Width: Use stereo imaging tools to widen or narrow device placement, improving separation and spatial definition.

5. Level Balancing

- Relative Levels: Adjust device and vocal tiers relative to each different to preserve a balanced mix:

- Vocals: Ensure vocals are intelligible and outstanding with out overshadowing instrumental nuances.

- Instruments: Balance instrumental elements to complement vocals and provide help without competing for interest.

6. Effects and Ambiance

- Reverb and Delay: Apply results judiciously to beautify spatial depth and cohesion:

- Vocals: Use diffused reverb to region vocals in a natural acoustic space, adjusting prepost pone and rot time for readability.

- Instruments: Apply results sparingly to create intensity and dimension, ensuring they mixture seamlessly with the vocal presence.

7. Monitor and Refine

Continuous Evaluation: Regularly pay attention severely to the combination, adjusting stages, EQ, and effects as needed to optimize readability and impact.

Reference Tracks: Compare your mix in opposition to professionally mixed tracks to gauge balance, clarity, and standard effect.

ACHIEVING READABILITY AND SEPARATION

Achieving readability and separation in a combination is essential for ensuring that every device and vocal can be heard noticeably and contributes correctly to the overall sound.

Techniques for Achieving Clarity and Separation

1. Frequency Management

EQ Carving:

- High by pass Filtering: Remove needless lowend frequencies from non bass gadgets and vocals to smooth up the mixture.

- Notch Filtering: Address frequency clashes with the aid of notching out specific frequencies in which gadgets or vocals overlap.
- Frequency Masking: Use EQ to carve out space for each element, making sure they occupy distinct frequency degrees without competing.

2. Dynamic Range Control

Compression:

- Vocals: Apply gentle compression to even out vocal dynamics, ensuring regular presence without sounding overly processed.
- Instruments: Use compression to manipulate dynamic peaks, allowing each tool to sit extra prominently within the mix without overwhelming others.

3. Spatial Placement

Panning and Stereo Imaging:

- Lead Elements: Place vocals and lead gadgets centrally or barely off middle to preserve attention and clarity.

- Instrumentation: Distribute units throughout the stereo area to create a sense of width and separation, ensuring every detail has its own sonic space.

4. Level Balancing

- Relative Levels: Adjust device and vocal stages relative to each different to hold a balanced blend:

- Lead Vocals: Ensure vocals are clean and outstanding, adjusting stages to sit down just above helping instrumentation.

- Supporting Instruments: Balance stages to offer help and texture with out overpowering lead elements.

5. Use of Effects

Reverb and Delay:

- Short Ambience: Apply subtle reverb or postpone to create a sense of intensity and area around vocals and units, improving separation.

- Tailored Effects: Use consequences with adjustable parameters (predelay, decay time) to in shape the context of every track, preserving readability at the same time as adding size.

6. Automation

Volume and EQ Automation: Automate volume tiers and EQ settings during the combination to control dynamic shifts and highlight crucial factors throughout one of a kind sections of the song.

Effects Sends: Automate outcomes sends to alter the amount of spatial processing implemented to vocals and gadgets, enhancing readability and separation in specific elements of the song.

7. Reference Mixing

Compare with Reference Tracks: Regularly reference your mix against professionally mixed tracks to evaluate readability, separation, and typical stability.

Critical Listening: Listen seriously throughout special playback systems (monitors, headphones, vehicle stereo) to make sure clarity and separation translate across diverse environments.

MONITOR MIXING

Monitor mixing is a critical aspect of stay sound reinforcement, ensuring that performers on stage pay attention themselves and other musicians honestly.

Monitor Mixing: Overview and Importance

1. Purpose

Personalized Audio Mix: Provides each performer with a tailor made mix in their vocals and contraptions, letting them pay attention themselves and others truly on stage.

Performance Confidence: Enables musicians to stay in sync, pitch, and rhythm, improving standard performance pleasant.

Feedback Prevention: Helps prevent remarks by using ensuring performers hear themselves effectively while not having to growth level monitor volumes excessively.

2. Components of Monitor Mixing

Monitor Speakers: On degree speakers or near monitors (IEMs) used to supply personalized mixes to performers.

Monitor Console or System: Dedicated mixer or console used by the monitor engineer to create and modify character mixes for every performer.

Communication: Clear communication among display engineer and performers to recognize their tracking needs at some stage in soundcheck and performance.

3. Key Considerations

- Mix Clarity and Balance: Ensure vocals are clear and outstanding, instrumental stability is suitable, and dynamics are controlled correctly.

- Feedback Management: Use graphic EQs or notch filters to save you remarks even as maintaining clarity in display mixes.

- Stage Layout: Position monitors strategically to avoid spillage into microphones and maintain separation between specific performers' mixes.

- Consistency: Maintain steady monitoring ranges all through the performance to hold performers comfortable and targeted.

4. Techniques and Best Practices

SOUNDCHECK: CONDUCT
THOROUGH
soundchecks to modify monitor mixes
primarily based on performers' options and
desires.

Communication: Establish clear
communication channels between
performers and display engineers for
realtime adjustments during rehearsals and
performances.

Monitor Mixing Tools: Utilize equipment
like mute companies, solo capabilities, and
cue sends to control and refine monitor
mixes effectively.

Monitoring Systems: Choose between
wedge monitors and inear video display
units (IEMs) based on performers' choices
and venue acoustics to optimize tracking
clarity.

5. Advantages of In Ear Monitoring (IEM)

Personalization: Offers unique manipulate over individual mixes, reducing level extent and improving clarity.

Isolation: Minimizes outside noise, permitting performers to focus on their mix without distractions.

Feedback Reduction: Less at risk of comments compared to traditional wedge monitors, particularly in difficult acoustic environments.

CREATING MIXES FOR PERFORMERS

Creating mixes for performers involves tailoring audio setups in order that each performer hears a balanced and clean blend of vocals and instruments that enables them carry out effectively on stage.

Steps to Create Monitor Mixes for Performers

1. Understanding Performers' Needs

Communication: Prior to soundcheck, speak with every performer to recognize their alternatives, which contraptions or vocals they want to pay attention prominently, and any specific changes they may require.

Reference Tracks: If feasible, pay attention to reference tracks or preceding performances with the performers to apprehend their perfect blend.

2. Setting Up Equipment

Monitor System: Determine whether or not you will use wedge video display units or inear video display units (IEMs) based totally at the venue, performers' possibilities, and acoustic issues.

Positioning: Place video display units strategically to decrease feedback and ensure most excellent sound insurance on level.

3. Soundcheck Process

Individual Mixes: Start by putting in place every performer's mix in my view on the monitor console.

Basic Levels: Set initial tiers for vocals and units based on soundcheck input and modifications.

Instrument Priorities: Prioritize lead vocals and critical instruments (e.G., rhythm segment for bands) that each performer wishes to pay attention genuinely.

5. Mix Adjustment

- EQ and Dynamics: Use EQ to carve out frequency area for every instrument and vocal, making sure clarity and stopping masking.

- Compression: Apply gentle compression to smooth out dynamics and make sure regular ranges with out overpowering other elements.

Effects Use: Add results like reverb and delay to enhance the spatial presence of vocals and gadgets, adjusting ranges to healthy the performance context.

5. Personalization and Fine Tuning

Individual Preferences: Fine tune mixes primarily based on each performer's remarks at some point of soundcheck, adjusting stages, EQ, and effects as wished.

Dynamic Changes: Anticipate and put together for dynamic changes for the duration of the performance, making changes to make certain performers continue to be comfortable and on top of things in their blend.

6. Monitoring During Performance

Real Time Adjustments: Remain aware of performers' needs all through the performance, making actual time adjustments to mixes primarily based on their requests or changing performance dynamics.

Feedback Prevention: Use photo EQs or notch filters to address capacity comments troubles without compromising blend clarity.

HANDLING FEEDBACK

Handling feedback, especially in stay sound environments, is critical for keeping clear and uninterrupted audio in the course of performances.

Strategies for Handling Feedback

1. Identify Potential Causes

Microphone Placement: Ensure microphones are positioned away from video display units and audio system to reduce the hazard of feedback loops.

Monitor Placement: Position level video display units strategically to avoid direct microphone pickup and remarks.

Room Acoustics: Be privy to reflective surfaces that could expand sure frequencies, probably causing comments.

2. Soundcheck Procedures

Gain Structure: Set appropriate benefit ranges for microphones and gadgets at some point of soundcheck to keep away from excessive amplification that leads to remarks.

Monitor Mixes: Ensure performers' reveal mixes are nicely balanced and clean, lowering the need to increase reveal volumes excessively.

3. Using EQ and Filters

Graphic EQs: Use image EQs to become aware of and reduce frequencies at risk of remarks, typically inside the mid range (800 Hz to 2 kHz).

Notch Filters: Apply notch filters to surgically dispose of unique frequencies in which feedback takes place, with out affecting the overall sound satisfactory.

5. Monitor System Management

Monitor Engineer Awareness: Have a dedicated screen engineer who video display units ranges and responds promptly to any feedback problems during the overall performance.

Communication: Maintain clear verbal exchange with performers to modify screen mixes quick in the event that they enjoy feedback.

5. Technique Adjustments

Microphone Technique: Instruct performers on right microphone method (e.G., proximity impact, angle of technique) to reduce comments risk.

Monitor Usage: Encourage performers to apply inear video display units (IEMs) in preference to degree monitors, as they're less prone to remarks because of their remoted nature.

6. Real Time Management

Monitor Levels: Monitor degree volume levels constantly at some stage in the overall performance, adjusting tiers as vital to prevent comments.

Immediate Response: Act quick if remarks occurs, the use of mute buttons or faders to attenuate problematic frequencies before they expand.

Identifying and doing away with feedback Identifying and casting off remarks in stay sound conditions requires a systematic approach and knowledge of the factors that make a contribution to remarks loops.

Identifying Feedback

1. Listen for Signs of Feedback:

High Pitched Tone: Typically inside the mid to excessive frequencies (normally around 2 kHz).

Sustained or Oscillating Sound: Continuous or fluctuating noise that will increase in intensity.

2. Monitor Equipment and Settings:

Use headphones or a monitoring device to listen for any onset of comments during soundcheck or performance.

Observe the picture EQ or spectrum analyzer for frequency peaks that indicate ability remarks frequencies.

3. Visual Inspection:

Check microphone and reveal placements visually to make sure they are positioned effectively to reduce comments.

ELIMINATING FEEDBAC

1. Adjust Microphone and Monitor Placement:

Microphone Positioning: Ensure microphones are positioned faraway from speakers and monitors to lessen direct sound pickup.

Monitor Placement: Angle monitors away from microphones and adjust their distance to minimize comments.

2. Manage Monitor Mixes:

Monitor Engineer Role: Assign a display engineer to actively manage and adjust reveal mixes all through the overall performance.

Balanced Mix: Ensure every performer's display mix is balanced without excessively excessive volumes that could lead to comments.

3. Use of EQ and Filters:

Graphic EQ: Use a picture EQ to discover and reduce frequencies which are prone to comments, commonly inside the 2 kHz range.

Notch Filters: Apply notch filters at the photo EQ to surgically cast off specific comments frequencies with out affecting the general sound nice.

4. Technique Adjustments:

Microphone Technique: Instruct performers to maintain steady microphone technique, averting surprising actions or setting the microphone too close to audio system.

Instrument Handling: Ensure instrument amplifiers are located away from microphones and monitor audio system.

5. Real Time Management:

Immediate Response: If comments takes place for the duration of the overall performance, decrease the microphone or tool quantity without delay.

Use of Mute Buttons: Use mute buttons on mixer channels or faders to fast attenuate the remarks frequency.

6. Feedback Suppression Devices:

Automatic Feedback Suppressors: Consider the usage of committed hardware or software program feedback suppressors which can locate and suppress comments frequencies in real time.

7. Room Acoustics and Environment:

Acoustic Treatment: Address acoustic reflections and room resonances which could exacerbate remarks problems.

Sound Check: Conduct a radical sound take a look at to pick out capacity feedback frequencies and adjust settings as a consequence.

8. Educational Approach:

Performer Education: Educate performers about feedback causes and prevention strategies to minimize occurrences all through performances.